Helmut Albert

2D ATOMS

Nuclear shapes and nuclear structure of atoms

Fundamental Physics

Contents

Imprint:

2D atoms. Nuclear shapes and nuclear structure of atoms
Translated from German: DeepL pro.

Texts, Cover: © 2023 Helmut M. Albert
© 2023 Helmut M. Albert, Talstraße 63; 79102 Freiburg. Germany

Herstellung und Verlag: BoD – Books on Demand, Norderstedt

ISBN: 9783757854119

2D ATOMS

Nuclear shapes and nuclear structure of atoms

Helmut M. Albert

Preliminary remarks

„A lot goes a long way" is not always true. In the case of fundamental physics it is even counterproductive. Many people like to be put off by the mass of terms, particles and theories when it comes to the atom. This is understandable! But - it is not the complexity of atoms and matter, but the complexity of science and scientists. The nature works according to simple laws!

Some text passages and passages in this paper refer to earlier publications of the author, which are listed in the bibliography(cf. Albert G./ Albert H. 2017; Albert 2019a; 2019b; 2021).

Helmut Albert, Freiburg/ Breisgau in March 2023.

1.0 View on the atomic nucleus

Which nuclear structure and nuclear shapes can be used to explain atoms conclusively? This question is addressed in the present work by contrasting the scientific theories of multiform atomic nuclei with the counterposition of an exclusively two-dimensional nuclear structure. To this end, the idea of spherical atomic nuclei in the first atomic models at the beginning of the 20th century is first described. Then, the alpha-cluster theory, which emerged soon after the discovery of the proton and the neutron, the nuclear shell theory, and the ideas of multiform atomic nuclei are pointed out.

Following this, we will look at selected research projects on the atomic nucleus from the last decade. In 2013, researchers confirmed for the first time that they could observe pear-shaped and other nuclear shapes in experiments in addition to spherical atomic nuclei. Spherical atomic nuclei are said to be atoms in the stable ground state, while the other nuclear shapes are to be regarded as deformations due to excitations or supernumerary nucleons. Further research projects on nuclear shapes and nuclear structure are then described.

The research reports reveal the scientists' efforts to reconcile the empirical research results with their ideas about nuclear structure. It can be seen that researchers often draw on previous theories in their interpretations. Regardless of the theories, however, it seems problematic to be able to explain the nuclear structure of atoms with the observed nuclear shapes. It is questioned by the author whether the external influences and excitations of the nuclei in particle accelerators are not too massive to draw conclusions about the natural basic shape of the atomic nuclei from the observations.

In order to show that the atomic structure can also be understood in a different way, an opposing position in the form of the checkerboard-like, planar atomic structure is contrasted. In each case, a research report is followed by an opposing position. A brief summary completes the picture of the checkerboard planar atomic structure. The starting point for the understanding of the whole atomic structure is then the properties of the atomic building blocks proton and neutron. Already the only possible spin settings, "spin-up" and "spin-down" of the nucleons, give a hint to the two-dimensional nuclear structure, which is explained in the thesis.

The present work makes no claim to completeness of the scientific ideas and theories about the atomic or nuclear structure.

2.0 Development of the atomic conception

At least since the beginning of the 20th century, the atomic nucleus has been considered as a spherical particle system with positive electric charge. The scientist J.J. Thomson(1856-1940), who presented his atomic model to the public in 1904, described the atom as "[...] a shpere of uniform positive electrification" (Thomson, 1904). The idea of the spherical shape of the atomic nucleus did not change when in 1911, Ernest Rutherford (1871-1937), based on his scattering experiments, rejected Thomson's atomic model and published the first nuclear-shell atomic model in history. In contrast to Thomson, Rutherford assumed that the electrons are located in an atomic shell, around the spherical atomic nucleus. His student Niels Bohr(1885-1962), who modified Rutherford's atomic model, said in 1922 about the atom: "This picture, at first sight, bears an extraordinary resemblance to a planetary system [...]"(Bohr, 1924: 6). Rutherford discovered the proton as the building block of the atomic nucleus in 1919 during the transformation of atomic nuclei, and a few years later, in 1932, James Chadwick (1891-1974) succeeded in identifying the neutron as the second building block of the atomic nucleus. Prompted by the discoveries, many researchers soon sought to decipher the structure of protons and neutrons in the atomic nucleus. Under the title The Alpha-Particle Model of the Nucleus, the physicists Lawrence R. Hafstad(1904-1993) and Edward Teller(1908-2003) published in 1938 their model conception of the structure of the atomic nucleus (cf. Hafstad and Teller, 1938). They assumed that atomic nuclei are built up of subunits of alpha particles. An alpha particle is a helium particle consisting of two protons and two neutrons. They thereby include the possibility that one proton or one neutron is additionally present. Likewise, in this model conception, a proton or neutron can be missing, which is then understood as a "hole" (cf. Hafstad and Teller, 1938).

Another theory of the atomic nucleus, which was published in the middle of the 20th century, is the nuclear shell theory, according to which an atomic nucleus has a shell structure similar to that of the atomic shell. Mayer-Kuckuk(1992: 194) writes in his standard work nuclear physics "The filling up of the nuclear shells occurs in a very similar way as the filling up of the electron shells in the shell". Certain proton and neutron numbers, form so-called magic numbers. According to the nuclear shell model, for example, the atomic nuclei of helium-4, oxygen-16, calcium-40 and nickel-56, have magic numbers for both protons and neutrons (cf. Mayer-Kuckuk 1992: 194).

Different nuclear shapes of atoms had already been known and described in the 1990s. Spherical atomic nuclei are according to Mayer-Kuckuk(1992: 193f.) nuclei with completely filled nuclear shells and are therefore stable nuclei. Similar to the drop of a liquid, an atomic nucleus can perform oscillations of the outer surfaces around its rest position. According to Mayer-Kuckuk(1992:213), an oscillating atomic nucleus will at first still preserve its nuclear shapes because of the "pairing forces". But if one increases the number of nucleons lying outside the shells, which can be called "valence nucleons"(ibid. 1992: 201), the possibility of deformation of the nucleus increases(cf. Mayer-Kuckuk 1992:213).

If much more nucleons are added, the position of the valence nucleons can change and the nucleus can deform into a rotating ellipsoid, making the nuclei more unstable(cf. Mayer-Kuckuk, 1992: 213f.). According to Mayer-Kuckuk (1992: 226), particularly many "cigar-shaped" deformations are found in the rare earths.

According to this, such elongated atomic nuclei, if they rotate very fast, can also assume other forms. As a model, one can imagine drop-shaped nuclei with properties of shells which, due to certain conditions, show stretched, flattened or multi-axial deformations of the nuclear shapes (cf. Mayer-Kuckuk, 1992: 227f.).

3.0 Explanation of terms

Atom and atomic nucleus

Doctrine: An atom (except hydrogen atom) consists of an atomic nucleus of electrically positively charged protons and neutral neutrons and an atomic shell in which electrons are located. The numerical ratio of electrons to protons determines the chemical properties of an atom. The atomic nucleus carries the largest mass of the atom, about 99.9%.

Checkerboard-like planar atomic structure: An atom(except hydrogen atom) is a binary system, built up of right-rotating protons and left-rotating neutrons. The proton-neutron configuration of an element determines its chemical properties. The total atomic mass is carried by the so called "atomic nucleus". Since there is no "atomic shell" afterwards, the terms atom and atomic nucleus are synonymous(cf. Albert G./ Albert H. 2017).

Nuclear shapes

Doctrine: spherical atomic nuclei are stable, while flat, elongated and other-shaped atomic nuclei are considered deformed and unstable.

Checkerboard-like planar atomic structure: Atoms, or atomic nuclei, build up planarly into atomic rectangles. An atomic rectangle whose building block rows are completely occupied represents a noble gas configuration. An atomic rectangle in whose building block rows not all nucleon sites are occupied is reactive(cf. Albert G./ Albert H. 2017).

Nuclear structure

Doctrine: According to the nuclear shell model, an atomic nucleus consists of shells of certain levels, similar to the electron shells. When these shells are completely filled with protons and neutrons, they are "magic nuclei". These nuclei have certain numbers for both protons and neutrons. However, they are not identical to the electron numbers of the noble gases.

Checkerboard-like planar atomic structure: Atoms build up in a checkerboard-planar fashion by protons and neutrons due to their opposite rotational directions. The nucleons are built up in a square lattice. Like in a gearbox, a right-rotating particle is directly followed by a left-rotating one in the rows of building blocks of the atoms. Diagonally, the same building blocks are opposite each other. The construction of the heavy atoms only takes place in two opposite directions The complete seven building block row of a heavy atom consists either of 3 protons and 4 neutrons, or of 4 protons and 3 neutrons.

Valence nucleons

Doctrine: "Valence nucleons" got their name according to Mayer-Kuckuk analogous to the so-called valence electrons. These nucleons are from today's conception outside of the closed nuclear shells and therefore contribute to the nuclear spin. They are supposed to be responsible for rotation states and deformations of atomic nuclei. (Mayer-Kuckuk 1992: 201).

Checkerboard-like planar atomic structure: According to the theory of checkerboard atomic structure, atoms build up by protons and neutrons in a square/rectangular fashion from the inside to the outside. Because of the equal sized atomic building blocks, already with 2x2 building blocks (2 protons, 2 neutrons) a closed atomic square is reached. Further neutrons are then outside this "frame", as in the case of the isotope helium-5. The same is true for all noble gases. As also with the atomic rectangle of neon which forms a closed atomic rectangle with 4x5 building blocks (10 protons, 10 neutrons). Each further proton or neutron is then outside of the neon atomic rectangle. If you like, these nucleons can be called valence nucleons. However, the "valence" or "value" is in reality not determined by them, but by the "gaps" or unoccupied nucleon sites in a building block series(cf. Albert G./ Albert H. 2017).

Spin up, spin down:

Doctrine: According to today's view, the so-called Pauli principle is valid for nucleons as well as for electrons. According to this, two neighboring particles must differ in one property (quantum number). This led to the fact that with the conception of several nucleons in a nucleus, for example two protons in the helium atom, must have two different spins. This means that in helium-4, one proton rotates right-handed and the other left-handed. The same is valid for the neutrons. Each type of building block can then have the two different spin settings.

Checkerboard-like planar atomic structure: The importance of spin in atomic structure cannot be overstated. Spins, angular momentum, or rotational directions are critical to atomic structure. Spin determines the identity of a nucleon as a proton or neutron. A proton is a right-rotating atomic component (as, for example, the Earth is a right-rotating planet), the neutron is a left-rotating atomic component. By transformation a proton becomes a neutron and vice versa a neutron becomes a proton. During these processes the rotation axis of a nucleon is reversed by about 180°. Therefore, a right-rotating atomic component is a proton and a counter-rotating atomic component is a neutron. See Figs. 1 and 2(cf. Albert G./ Albert H. 2017)(Albert H. 2019a).

Strong nuclear force

Doctrine: the strong nuclear force is explained by science with the so-called Quantum chromodynamics(QCD). According to this, the cohesion in the nucleus is mainly traceable to the subparticles of the nucleons, the quarks and gluons. In addition, a kind of "residual interaction" is supposed to be effective among the nucleons.

Checkerboard-like planar atomic structure: Attraction and repulsion of protons and neutrons form the strong nuclear force of the atomic nucleus. Atomic components with the same direction of rotation repel each other, while atomic components with opposite direction of rotation attract each other(cf. Albert G./ Albert H. 2017: 12).

In contrast to the previous theory that only protons repel each other, researchers were able to prove that neutrons also repel each other(cf. Schmidt et al. 2020). This confirms the theory of the checkerboard-like atomic structure by protons and neutrons. Despite the form-stable atomic structure, only oppositely rotating building blocks touch each other. In this way, the atomic nuclei build up like gears, along a row of building blocks alternating one proton, one neutron. In figure 1, the points of contact of the nucleons in the atomic nucleus are shown.

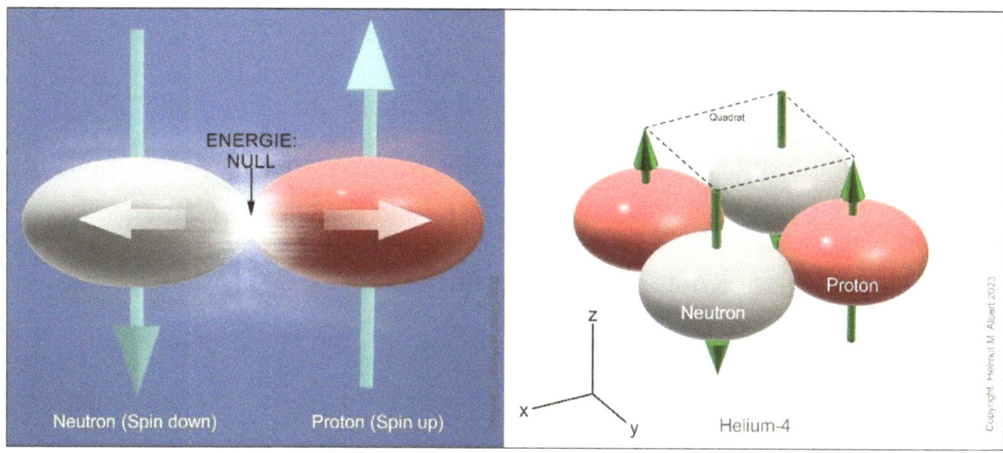

Fig. 1. Proton-neutron. Fig. 2. Checkerboard-like planar atomic structure; Helium-4 isotope (angular momentum vectors).© 2023 Helmut M.Albert

4.0 Deformed atomic nuclei

After the year 2000, more efforts were made to experimentally study multiform atomic nuclei and their nuclear structure. For example, an international team of researchers led by Peter Butler(2013) was able to detect pear-shaped deformed atomic nuclei at the CERN research center. In Nature, the researchers report how they extracted the isotopes radon-220 and radium-224 from primordial nuclei and then excited them by bombarding them with other targets. In this experimental study, they found that the nuclei exhibited octupole deformations and took on the shape of a pear. For the isotope radium-226, the researchers predict an even stronger pear-shaped deformation of the nucleus (see Gaffney et al., 2013).

The researchers refer to an atomic nucleus as a "quantum system"(Gaffney et al., 2013) whose nuclear shapes depend on the nucleons and their nuclear forces. Atomic nuclei whose shells are completely filled with protons and neutrons are called doubly magic. Such spherical nuclei are considered stable atomic nuclei, the researchers said.

Excitations of these spherical nuclei or the addition of nucleons deform the atomic nuclei. Most often, the researchers said, this results in flattened or elongated nuclear shapes. Less commonly, spherical atomic nuclei deform by what is known as octupole deformation (Gaffney et al., 2013), which could be important in the search for dipole properties of nuclei.

The complexity and high mass number of an atomic nucleus complicate the calculations, the researchers said. However, the experiments could show that heavy nuclei, especially the radium-224 nucleus, are excited by collisions, deform, and become unstable (see Gaffney et al., 2013).

4.1 Counterposition

Radon, like all noble gases, is non-reactive, or only exceptionally so. Radon isotopes are associated with more or less rapid natural decay processes from Rn-222 to Rn-218. Radon is a so-called alpha emitter ("Radon" 2023). When artificially excited, radon isotopes radiate even more strongly. Such nuclear radiation is the radiation of an excited or decaying nucleus, whose radiation is spherical. Whether therefore the nucleus itself is spherical is another question. If such atomic nuclei are also used as projectiles or targets, it stands to reason that a variety of phenomena and appearances may occur, depending on the collision. However, it is very questionable whether a reliable statement about the nuclear shapes can be derived from this. In particular, it seems hardly possible to deduce the nuclear shapes of an atom from the experimental results.

5.0 Different nuclear shapes

The importance that fundamental physicists attach to the observations of nuclei is evident in many research reports. According to the Canadian research team led by Paul Garrett(2019), nuclear shapes are an important aspect by which nuclear structures can be experienced. In addition to spherical nuclei, the researchers were able to observe complicated, flat, elongated, and ellipsoidal nuclei. These atomic nuclei can take on a variety of shapes, the researchers said. Two nuclear shapes in particular are seen more often next to each other, which used to be rare. These are spherical and more deformed atomic nuclei, which are associated with rotations(cf. ibid. 2019).

The starting point always seems to be the spherical ground state of an atomic nucleus. Here, the deformations of the nuclei start from a "proton-neutron-quadrupole interaction". Ground state nuclear shapes always have closed nuclear shells, meaning that there are no valence nucleons(Garrett et al. 2019). Using the cadmium isotopes, Cd-110 and Cd-112, experiments were able to demonstrate the "coexistence"(ibid. 2019) of two nuclear shapes. The researchers were able to gather further information by isotopic spectroscopy of lead isotopes In the process, they discovered more than two nuclear shapes in lead isotopes. When excited appropriately, both spherical, flat, and elongated nuclear shapes were revealed. Figure 3 illustrates the findings the research team gained from their investigations. Based on their observations, they assume that in lead isotopes, several ring shapes are found side by side in the nuclei (cf. ibid. 2019).

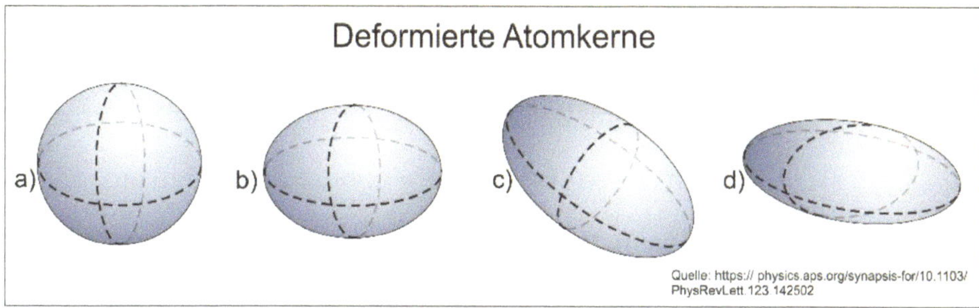

Fig. 3. Deformed atomic nuclei, drawing after: Garrett et al. 2019.

5.1. counterposition

Spherical or sphere-like nuclear shapes of stable atomic nuclei are in reality impossible. This is equally true for the nuclear shapes of cadmium isotopes. Conceivable are sphere-like oscillation or rotation phenomena of atomic nuclei, similar to a flat coin which is put into rapid rotation. Most likely, however, it is not so much the cadmium isotopes that are seen as the electromagnetic spherical radiations of particles.

A strong plastic change of isotopes due to a different neutron number is equally unlikely, unless by strong external effects on an isotope. Of course, cadmium isotopes, due to one or two more neutrons, show different behavior and values, but the basic proton configuration remains the same. That is, the basic plastic shape of an atomic nucleus does not change in the same way because of this. Figure 4, shows the checkerboard-planar configuration of the isotopes Cd-110 and Cd-112. The mass number of an isotope can be artificially reduced to the point where the checkerboard integration of protons dissipates. Naturally, an isotope decays, or transforms, when the energetic balance between protons and neutrons is no longer present. Protons have slightly more energy than neutrons. Let us consider the differences using the isotopes carbon-12 and carbon-13 as an example. While carbon-12 has equal numbers of protons and neutrons, carbon-13 has one neutron more. On the one hand, this changes the value, but on the other hand, C-13 then has a nuclear spin that C-12 does not have. Nevertheless, the configuration of the two isotopes remains the same except for one neutron difference. Their plastic nuclear shapes do not change immediately. #

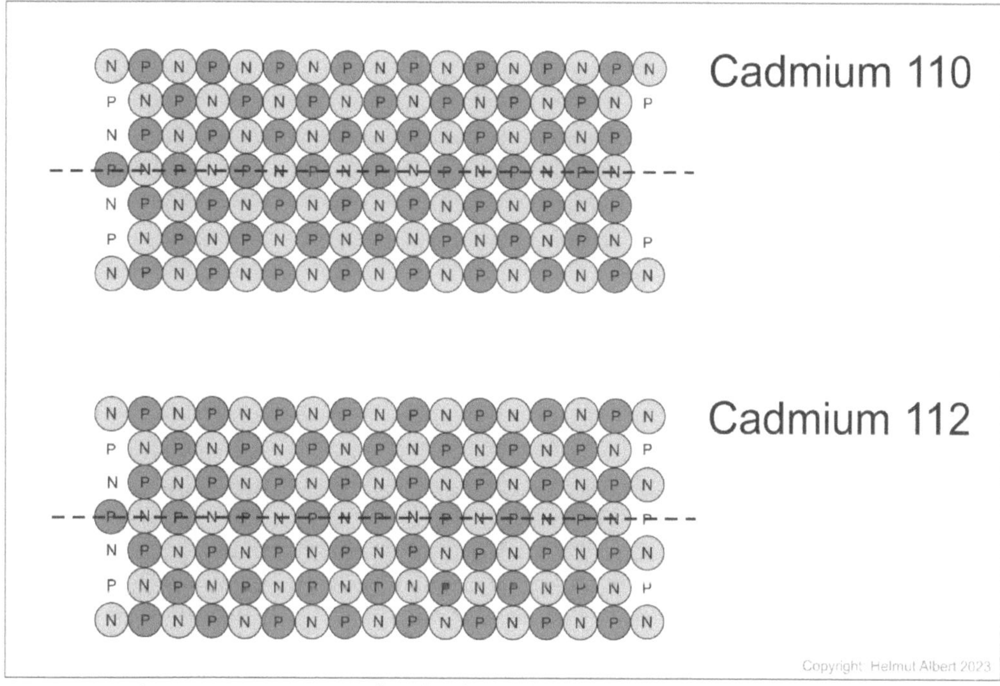

Fig. 4. Cadmium isotopes. Checkerboard-like planar atomic structure; © 2023 Helmut M. Albert

If we look at the proton-neutron configuration of the isotopes Cd-110 and Cd-112 on the basis of Figure 4, we notice the axially symmetrical arrangement of the isotopes, as is common in nuclei with even mass numbers. This is the reason that these isotopes do not have a nuclear spin, while Cd-113, for example, has a nuclear spin. The proportion ratio of cadmium isotopes is about 1: 2.3. The smallest cadmium isotope has sixteen rows of seven according to the checkerboard planar atomic structure. Like many heavy atomic nuclei, the stable Cd isotopes have more neutrons than protons. In this case, the two opposite end rows of the Cd-110 and Cd-112 isotopes are incomplete. According to the model conception of the cadmium nuclei shown here, it can be assumed that cadmium nuclei have 54 protons (atomic number 54) and not 48 as assumed by science so far. The arrangement of neutrons is also associated with the values of an atom, which change depending on the isotope. In Figure 5, the Cd isotope-112 is shown in perspective view with representation of the spins or rotational axes. This makes it easy to understand the checkerboard and planar structure of the Cd isotope by protons and neutrons. As in a gear, the oppositely rotating atomic building blocks, proton and neutron, alternate along the rows of building blocks.

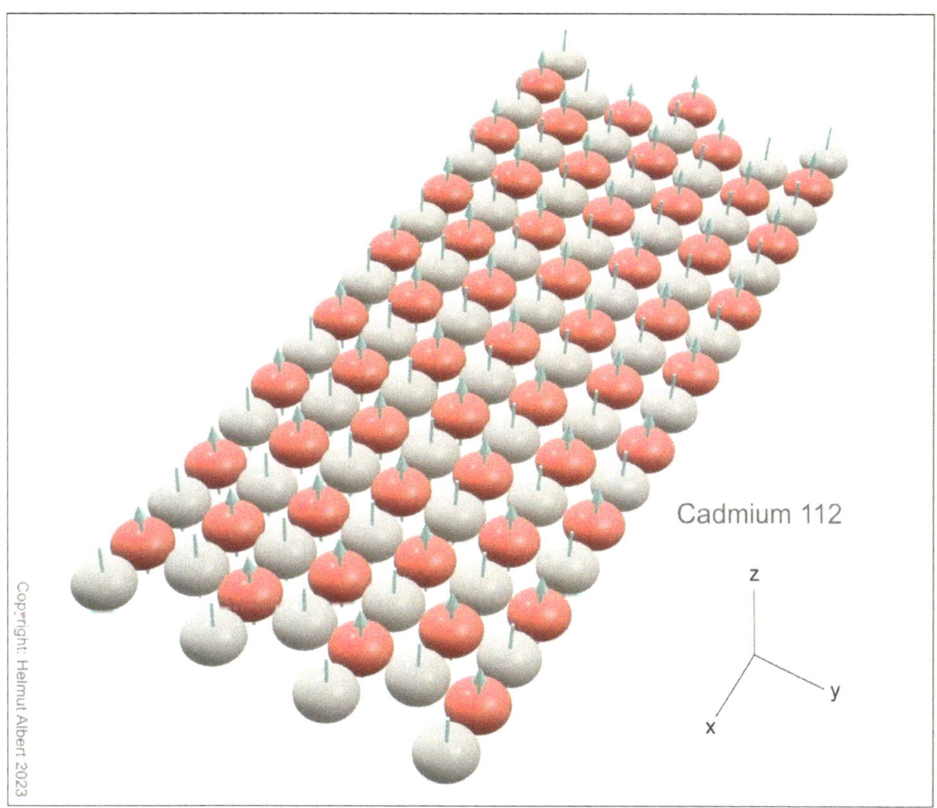

Fig. 5. Isotop, Cadmium-112. Checkerboard-like planar atomic structure; perspective view. (proton= red, neutron= gray. spin= green). © 2023 Helmut M. Albert

6.0 Nuclear structures of atomic nuclei

Also many other scientists try to explain the nuclear structures of atoms on the basis of observed nuclear shapes. Thereby it becomes clear that the theories from the middle of the 20th century still play an important role. For example, the theory that atomic nuclei are built up from subunits of alpha particles (clustering) (cf. Hafstad and Teller 1938). This idea of alpha particles forming clusters in atomic nuclei was addressed by a team of researchers led by Martin Freer (2018). They investigated the clustering of light atomic nuclei. They first addressed the question of how nucleons cluster together to form alpha particles. These alpha particles, once formed by nucleons, are expected to cluster together to form larger atomic nuclei, taking into account the Pauli principle. Based on clusters, an overall picture of atomic structure under the influence of the strong nuclear force will be drawn. In one illustration, the researchers show how they envision a tetrahedral (left) and a flat (right) arrangement of alpha particles (Helium-4) in the oxygen isotope O-16. See Figure 6, where the protons are marked in red and the neutrons in blue. According to this, protons should have both spin up and spin down, as well as neutrons(cf. Freer et.al 2018). Both spin settings are then possible for each of the two types of building blocks and are based on the Pauli principle. In this representation, two protons and two neutrons are combined to form one helium 4-particle each. According to the researchers, knowledge of clustering represents an important support for nuclear structure theory because it appears to explain the connection between light nuclei and the strong nuclear force. According to the researchers' idea, in this way, unions of two or more helium-4 particles to other atomic nuclei are possible (see ibid. 2018).

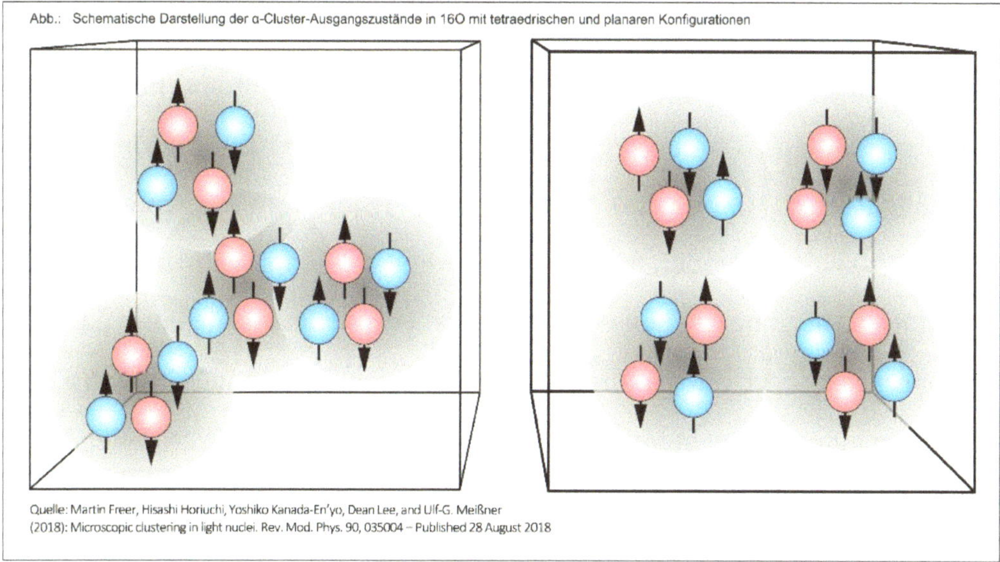

Fig. 6. Cluster formation; drawing after: Freer et.al. 2018)

6.1. Counterposition

If we evaluate the above considerations according to the checkerboard-planar atomic structure model, then the clustering described and shown above as a stable nuclear structure is not realistic in reality. Perhaps a clustering of helium particles as a transition state is possible in fractions of a second, but nothing more. But even then it is questionable whether they should be He-4 isotopes or rather He-3 or He-5.

In general, atomic nucleus with certain proton number(atomic number) can be considered as a certain element only as long as the proton-neutron configuration is intact. As with the nuclear fission from an element with a certain proton number two or more other elements, with other proton numbers become, it behaves also vice versa. Several helium particles approaching each other do not yet result in a new atomic nucleus with a new proton number. The atomic number 2 of the helium particles would have to remain after all, if one considers the idea of alpha particle clusters. Atomic nuclei with atomic numbers 4, 6 or 8 do not automatically result. Moreover, it must be considered that the proton-neutron configuration of helium-4 forms a closed system and He-4 nuclei therefore appear as non-reactive single atoms in nature. In contrast, the helium isotopes He-3 and He-5 are not closed systems. From this point of view, a merger to form the nucleus of beryllium-8 would be more likely. One cannot assume that heavier atomic nuclei come into being only under great pressure in stars.

According to the alpha particle cluster theory, helium-4 particles assemble to form other atoms(cf. Hafstad and Teller 1938). The illustrations of the research team around Martin Freer show helium-4 particles which are in a certain proximity to each other. See figure 6, but it is not clear which forces hold the noble gas particles together, forces like in molecules can not be! As already stated, the noble gases in nature are not reactive and therefore monatomic. Nevertheless, according to the idea of the researchers, from two helium-4-particles the element beryllium-8, from three helium-4-particles, the element carbon-12 etc. should be formed. But this is extremely improbable.

From the point of view of the checkerboard-planar atomic structure, it is impossible for the oxygen isotope-16 to have a configuration like that shown in Fig. 6. Also, a transition state of oxygen isotope-16 that lasts only fractions of a second is impossible in this form. Instead, all oxygen isotopes build up in a checkerboard-planar fashion. See Figure 7.

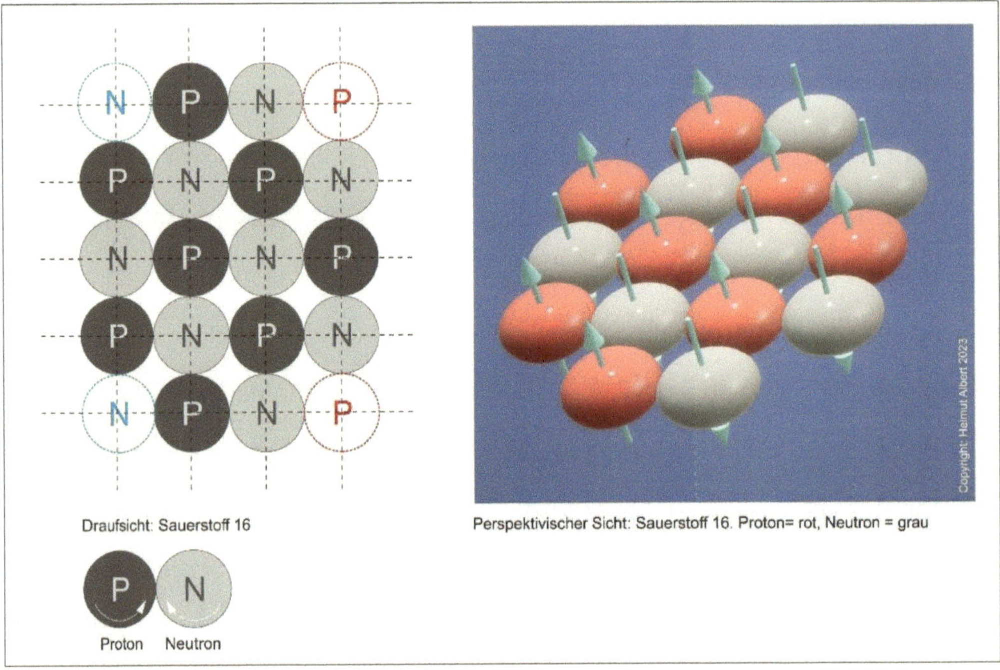

Fig. 7. Oxygen isotope-16. Checkerboard-like, nuclear structure. © 2023 Helmut M. Albert

A tetrahedral arrangement, as known from molecular geometry, is completely unthinkable for the protons and neutrons of an oxygen isotope. The representation of the spin of the protons and neutrons in figure 6 according to the theory of the scientists seems to be revealing. Here it becomes clear how an abstract rule like the Pauli principle leads to misinterpretations. The spins of the protons shown in illustration point downward as well as upward, just like the neutrons. According to this, the spin of a proton or neutron would be arbitrary or exchangeable, respectively, without changing the building block type. However, the representation of the spins does not correspond in any way to the actual conditions in the nucleus of an oxygen atom. An oxygen atom is an atom whose sixteen nucleons are arranged exactly checkerboard-like in a rectangle measuring 4 x 5 building block fields. Since all protons have spin up and all neutrons have spin down, the structure of the atom is like that of a gear.

7.0 Clustering of alpha particles

A similar idea of the nuclear structure, as the researchers before, has researchers around Otsuka T. (2022). They also assume that alpha particles form the nuclear structure. In their research report they describe experimental investigations, computer simulations and theoretical considerations.(cf. Otsuka, T., Abe, T., Yoshida, T. et al. 2022) According to an interpretation of the researchers, the isotopes beryllium-8 and carbon-12 consist of clusters of alpha particles (helium-4). Depending on the state of the atomic nucleus, the alpha particles are arranged in the atomic nucleus. According to the researchers, this could explain the different atomic nucleus shapes (cf. ibid. 2022). The researchers' illustrations show groupings of alpha particles. In Figure 8, called the cluster model of carbon-12, three different groupings of alpha particles are shown.

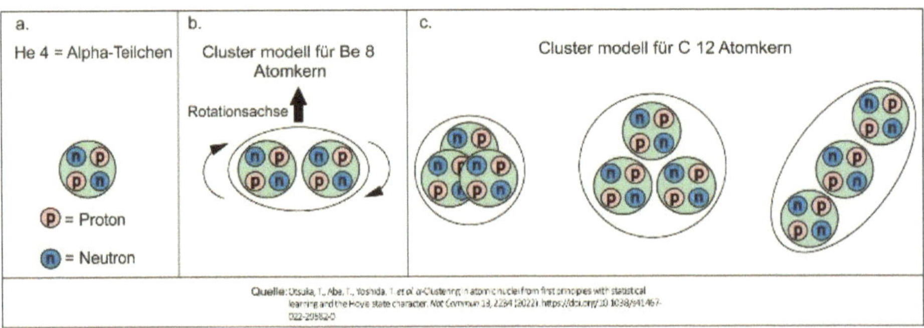

Fig. 8. Alpha particle cluster. Drawing after: Otsuka, T., Abe, T., Yoshida, T. et al. 2022).

7.1. Counterposition

If one considers this report on the nuclear structure according to the checkerboard-planar atomic structure, then one must evaluate the ideas of the researchers as fanciful, however without realistic basis. The atomic structure of beryllium and carbon nuclei(see Fig. 8) is impossible as pointed out by the researchers. What force should determine the different distances of nucleons and alpha particles from each other? Protons and neutrons cannot act in subunits of atomic nuclei. Since only the strong nuclear force is known as an attractive force in the atomic nucleus, there should be a staggering of the nuclear forces in an atomic nucleus. It would have to bind the nucleons stronger to each other, while the subunits would be bound weaker to each other. But a connection of He4-particles seems to be nonsensical already because of their known monatomicity.

If one looks at the representation of a single helium nucleus in fig. 2, then at least the arrangement of protons and neutrons almost corresponds to reality. Identical building blocks lie diagonally opposite each other in the atomic nucleus, while they alternate in a building block row. However, it is wrong that the protons and neutrons do not touch each other at their equators. But how else should the nucleons balance their opposite rotational direction.

Let us hold, an accumulation of two, three or four helium-4 ponds does not result in new atomic nuclei with new atomic number. The short-range nuclear force of the nucleons, which acts on all protons and neutrons, already speaks against this. The nuclear force is not exhausted by the fact that two protons and two neutrons combine with each other. In the same way, three or more protons and neutrons can combine in the square lattice.

8.0 Transition states

In an article by researchers Ebran and Kahn(2022: 4 - 15) in Spektrum-kompakt, the idea of alpha particles as subunits of atomic nuclei is also addressed. The researchers report on their concept of providing a common basis for assessing different atomic nuclei states. To this end, the researchers propose to classify the various states of atomic nuclei like the states of matter, solid, liquid, and gas. (cf. Ebran and Kahn et.al.2020:). According to Ebran and Kahn, clusters of helium particles (alpha particles) could thus be put on a common footing. In this they see a possibility to classify a multiplicity of made observations of nuclear states and effects. The scientists Ebran and Kahn refer to the theory of clustering of alpha particles in atomic nuclei, which was already established in the 1930s(cf. ibid. 2020: 13).

Against this background, they assume that the isotopes beryllium 8, carbon 12, oxygen 16 and neon 20 are structured by alpha particles (cf. Ebran, Kahn 2020: 13). Already in an earlier study on atomic nuclei, Kahn (2014) argued that, due to the clustering of helium particles, molecular structures are also possible in which the excess neutrons act as links between the helium particles.

Researchers Ebran and Kahn assume that protons and neutrons can arrange themselves in atomic nuclei in a similar way as atoms in molecules due to the strong nuclear force(cf. Ebran, Kahn 2020: 12).

Already in the 1950s, according to the researchers, the so-called "Hoyle state" (Ebran, Kahn 2020: 8-10) was described. This involves the arrangement of three alpha particles(helium-4) into a carbon-12 isotope.This process is also called the "triple-a process". The first step is the unification of two alpha particles to beryllium-8. In the next step Be-8 unifies with another alpha particle to carbon-12. But since Be-8 decays very fast this process seems improbable. Therefore, according to the "Hoyle state", the carbon isotope-12 should be able to exist in a state which combines three alpha particles for a time. Such a state could be proved, however, it was unstable (see ibid. 2020: 8-10). The traced figure no. 6 illustrates the idea of the so-called Hoyle state.

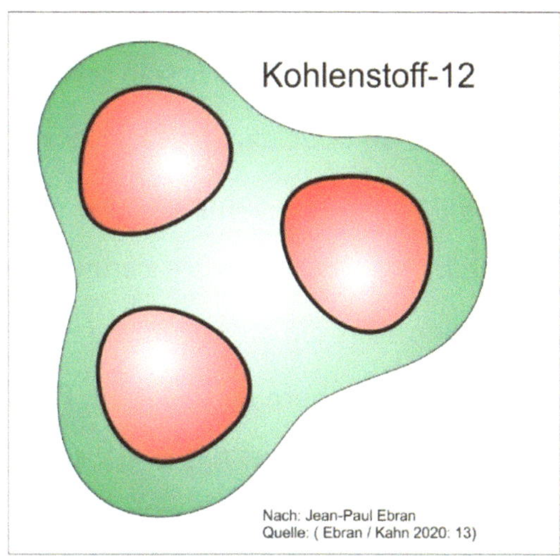

Fig. 9. Carbon-12, excited state. Drawing
after Jean-Paul Ebran. Source: Ebran /Kahn 2020.

8.1 Counterposition

The idea of aggregate states of atomic nuclei seems helpful at first in relation to the current scientific ideas of nuclear shapes and nuclear structure. But if we look at this idea realistically, it represents a further complication, with which real conditions cannot be explained. An overall picture cannot be achieved with it. Of course there are transition states in atomic nuclei, as in all particle systems, which form and dissolve again. There are rotational and vibrational states of atomic nuclei and many other states. But these states obscure the real nuclear structure and the real nuclear shapes of atoms. One should not give too much weight to the technical possibilities in this case.

The basis for the assumption of aggregate states in atomic nuclei is the idea that nucleons are built up of quarks and gluons. However, this idea is absurd and there is no real evidence for it. Even the term "gluons" (gluing) shows that nothing has been understood of the forces acting in the nucleus. To explain the nuclear structure, fundamental physicists like to resort to well-known molecular arrangements, which are transferred to the atomic level of nucleons. But nucleons are elementary particles! They exist because of their rotational energy which can be neither produced nor destroyed.

If one looks at the example of the carbon isotope-12, how such nuclei are built up, then ideas like the so-called "Hoyle state" are not very helpful. One has to give credit to Ebran and Kahn that they also pointed out the instability of such a state(cf. Ebran, Kahn 2020: 14).

A collection of helium nuclei cannot be understood as a carbon nucleus, since a fusion of the three helium-4 nuclei has not taken place and so cannot take place? Unless they would be completely shattered as a particle system due to strong external impacts and then rearrange themselves, but this is unlikely. Very probable is the idea that just the rarer isotopes helium-3 and helium-5 form heavier atomic nuclei. The same is true for the isotopes of other elements. Figures 10 and 11 show helium isotopes and carbon isotope 12 according to the checkerboard planar atomic structure. For the isotopes, the values are also shown.

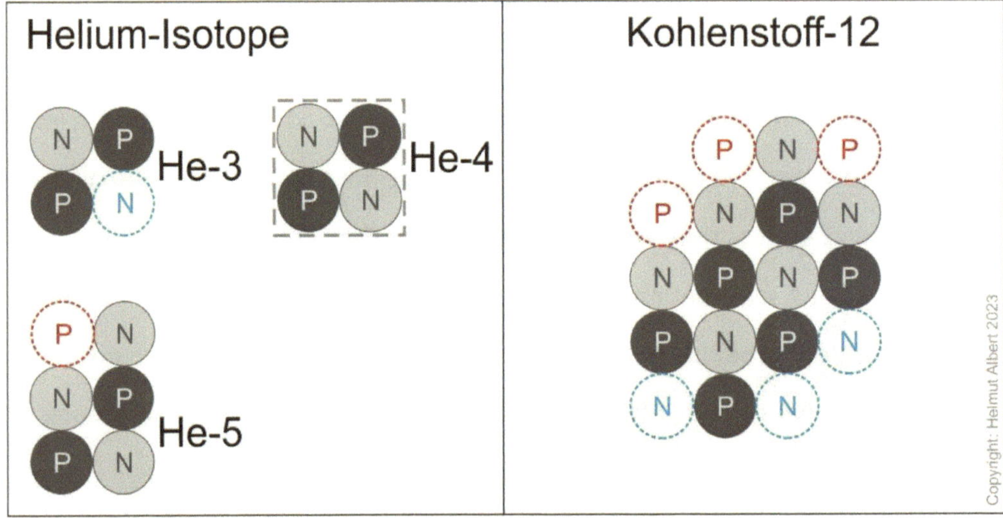

Fig. 10. Helium isotopes. Fig. 11. Carbon-12, Checkerboard-like atomic structure. © 2023 Helmut M. Albert

9.0 Appearance and reality

As can be read on the homepage of the TU Darmstadt, physicists of the TU Darmstadt together with Japanese physicists have carried out a project to investigate the deformation of atomic nuclei. It has long been known that there are nuclear shapes other than spherical nuclei, but why this is so was unclear for a long time (cf. Kremer et al. 2016).

In the project, the Japanese physicists related the forces of deformation directly to the protons and neutrons of an atomic nucleus. On this basis, they attempted to use computer simulations to reveal different nuclear shapes. They clarified that the zirconium isotope-96 is lighter in the spherical state than in the elongated deformed state. When the results of the Japanese physicists were empirically verified, the physicists at TU Darmstadt were able to confirm the results. This should make it possible in the future to calculate and predict the nuclear shapes of certain atomic nuclei The research report on this appeared in Physical Review Letters (cf. Kremer et al. 2016). There are numerous research projects like the one described above. Scientists try to create a theoretical basis for the calculation of atomic nuclei based on empirical results, or vice versa a theoretical basis is tested empirically. However, the question arises whether the observation of nuclear shapes can serve as a basis for the investigation of atomic nuclei at all, or whether such observations are not based too much on oscillation, rotation or radiation effects.

Can this be used to represent the actual nuclear shapes? Can the electromagnetic radiation, distort the nuclear shapes? Even if states of a nucleus can be roughly imaged with it, it is not the "direct" nuclear shell of an atom that is observed. The confirmation of spherical or sphere-like nuclear shapes by detection instruments fits well into the doctrine of the spherical atomic nucleus, but cannot convince. Already the eminent physicist Werner Heisenberg(1979), who was critical of observations of particles and their interpretation, expressed himself in this context as follows:

> „Sehen Sie, die Beobachtung ist ja im allgemeinen ein sehr komplizierter Prozess. Der Vorgang, der beobachtet werden soll, ruft irgendwelche Geschehnisse in unserem Meßapparat hervor. Als Folge davon laufen dann in diesem Apparat weitere Vorgänge ab, die schließlich auf Umwegen den sinnlichen Eindruck und die Fixierung des Ergebnisses in unserem Bewusstsein bewirken."(Heisenberg 1979: 31).

In the many research reports, the experimental results confirm by and large the theoretical ideas of fundamental physics, as they can also be read in the technical literature. Especially when it comes to spherical and deformed nuclear shapes. From the nuclear shapes observed by means of the most modern technology nuclear structures can be derived, so the idea of some physicists. With the theoretical nuclear structures then the previous idea of the basic structure of the nuclei is explained. But what amounts to a circular argument! For the science the question must arise, whether these detection devices can deliver reliable pictures of nuclear shapes with which the nuclear structure can be concluded? Technology can also deceive - a certain critical distance seems to be reasonable. A critical examination of fundamental physics was provided by the physicist Sabine Hossenfelder(2018) with her article Das hässliche Universum (The ugly universe). In it, Hossenfelder writes the following about the processes involved in the study of particles in particle accelerators:

> „Anders als Lichtmikroskope, die Spiegel und Linsen verwenden, nutzen Teilchenbeschleuniger elektrische und magnetische Felder, um Strahlen aus elektrisch geladenen Teilchen zu beschleunigen und zu fokussieren. Aber indem man Teilchen immer stärker beschleunigt, um einen Gegenstand zu untersuchen, wird es zunehmend schwieriger, aus der Messung Informationen zu

entnehmen. Das liegt daran, dass die Teilchen, die das Objekt untersuchen sollen, dieses zu verändern beginnen"(Hossenfelder 2018: 72).

In the opinion of the author of this paper, Hossenfelder thus shows how problematic it is to make reliable statements about microscopic observables with particle accelerators. Questions to what extent reliable images of atomic nuclei are possible at all cannot be considered a minor matter? Is it possibly a matter of radiation radiating spherically from bodies, even if a body is rectangular like a smartphone? These questions will be at least as important to science in the future as other questions about the atomic nucleus. Of course, a fast oscillating or rotating body can have a completely different appearance than at rest! To see this you only have to put a coin into fast rotation. The following figure shows a comparison of the observations of nuclear shapes scientific observations(cf. Mayer-Kuckuk 1992: 228) and the nuclear structure according to the checkerboard-planar atomic structure.

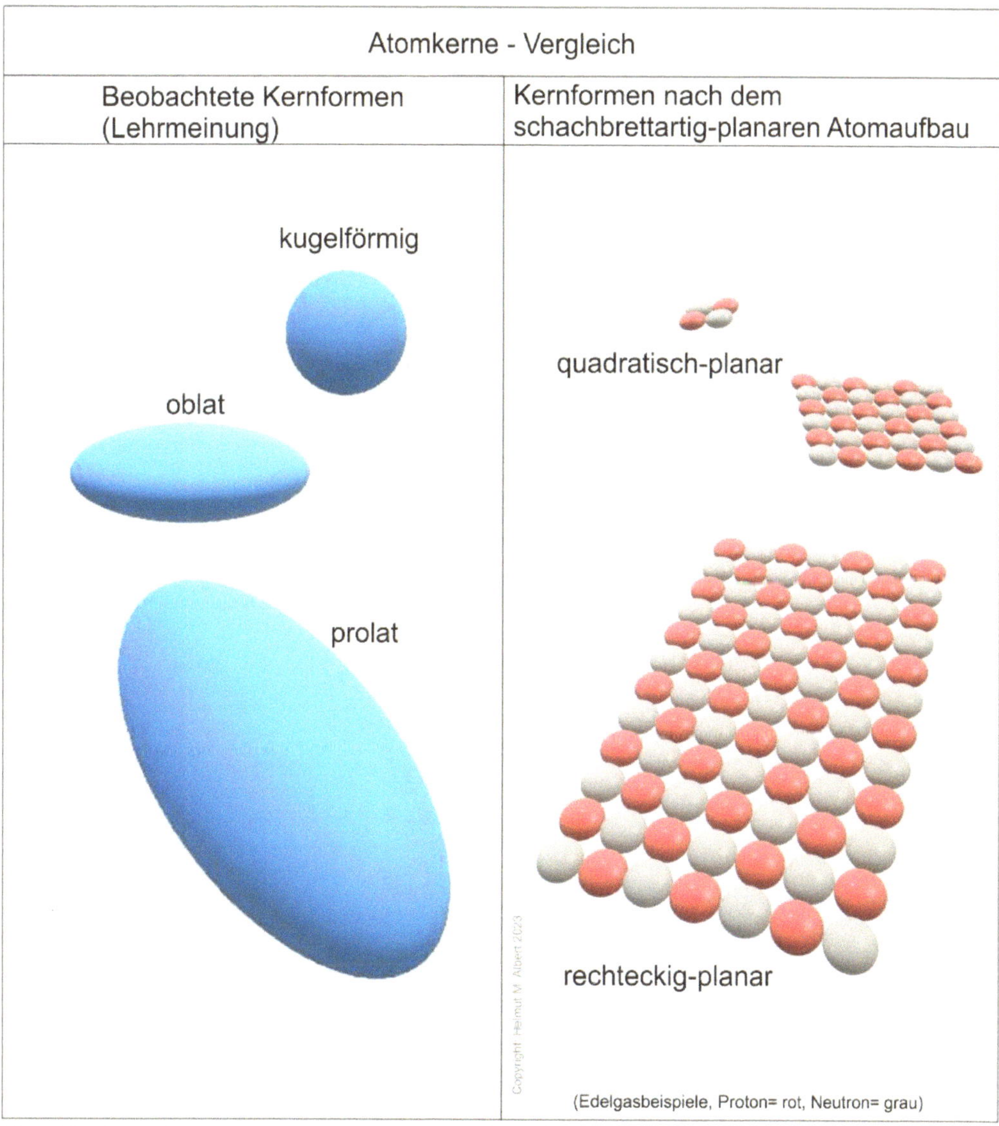

Fig. 12. Atomic nuclei - comparison. Left: Observed nuclear shapes. Right: Nuclear shapes. Checkerboard-like planar atomic structure. © 2023 Helmut M. Albert

10.0 List of figures

1. proton-neutron

2. helium-4; checkerboard-planar atomic structure

3. deformed atomic nuclei,
Drawing after: Garrett et al. 2019

4. cadmium isotopes, Cd 110, Cd 112

5. isotope Cd-112

6. cluster formation;
Drawing after: Freer et al. 2018

7. Oxygen isotope-16

8. alpha particle cluster;
Drawing after: Otsuka T., Abe T., Yoshida T., et al. 2018

9. C-12, excited state;
Drawing after: Jean-Paul Ebran; Ebran/Kahn 2020.pp. 10 and 12.

10. helium isotopes

11. carbon-12; checkerboard planar structure.

12. atomic model comparison

11.0 Bibliography

Albert, Helmut/ Georg, Albert(2017): Atomic model with checkerboard-like structure, (ed.) Verlag Helmut Albert, Freiburg. Epubli -Printing, Berlin.

Albert, Helmut (2019a): The spin nuclear force; Planar atomic model; Verlag Helmut Albert, Freiburg. Epubli -printing, Berlin.

Albert , Helmut (2019b): The binary system of atoms. Verlag Helmut Albert, Freiburg. Epubli -Printing, Berlin.

Albert , Helmut (2021): Imagine checkerboard atoms! [E-Book] Publisher Helmut Albert, Freiburg. Epubli , Berlin.

Bohr, Niels (1924): Über den Bau der Atome, Berlin, Verlag von Julius Springer 1924. lecture at the acceptance of the Nobel Prize in Stockholm 1922. translation: W. Pauli jr.

Ebran, Jean-Paul/ Elias, Khan(2020): Between liquid and crystal. The atomic nucleus. In: Spectrum of Science, compact. 25.20. [E-Book] Publication date: 22.06.2020. Verlagsgesellschaft mbH, Heidelberg. S. 5 - 14.

Ebran Jean-Paul/ E., Khan/ T. Nikšić/ D. Vretenar(2014): Density Functional Theory Studies of Cluster States in Nuclei. In: https://journals.aps.org/prc/abstract/10.1103/PhysRevC.90.054329, Published 20 November 2014

Freer, Martin/ Hisashi, Horiuchi/ En'yo Yoshiko, Canada/ Dean Lee/ Ulf -G. Meissner (2018): Rev. Mod. Phys. 90, 035004 - Published 28 August 2018.

Gaffney, L.P. / P.A. Butler/ M. Scheck/ A.B. Hayes/ F. Wenander/ M. Albers/ B. Bastin/ C. Bauer/ A. Blazhev/ S. Bönig/ N. Bree/ J. Cederkäll/ T. Chupp/ D. Cline/ T.E. Cocolios/ T. Davinson/ H. De Witte/ J. Diriken/ T. Grahn(...) M. Zielinska(2013): Studies of pear-shaped nuclei using accelerated radioactive beams. Nature 497, 199-204 (2013). https://doi.org/10.1038/nature12073

Garrett; Paul E./ T.R. Rodríguez/ A. Diaz Varela/ K.L. Green/ J. Bangay/ A. Finlay/ R.A.E. Austin/ G.C. Ball/ D.S. Bandyopadhyay/ V. Bildstein/ S. Colosimo/ D.S. Cross/ G.A. Demand/ P. Finlay/ A.B. Garnsworthy/ G.F. Grinyer/ G. Hackman/ B. Jigmeddorj/ J. Jolie (...)S.W. Yates(2019): Multiple shape coexistence in 110,112Cd. Phys. Rev. Lett. 123, 142502 - Published 3 October 2019. https://journals.aps.org/prl/abstract/10.1103/PhysRevLett.123.142502

Hafstad, Lawrence R./ Edward, Teller (1938): The Alpha-Particle Model of the Nucleus. DOI: https://journals.aps.org/pr/abstract/10.1103/PhysRev.54.681 Published 1 November 1938.

Heisenberg, Werner (1979): Quantum theory and philosphy. Reclam. Editor: Jürgen Busche. Reclam's Universal Library. No. 9948. p. 31

> *"You see, observation is in general a very complicated process. The process, which is to be observed, causes some events in our measuring apparatus. As a consequence of this, further processes then take place in this apparatus, which finally cause, in a roundabout way, the sensual impression and the fixation of the result in our consciousness.* (Translate: DeepL pro).

Hossenfelder, Sabine (2018): The ugly universe: why our search for beauty leads physics to a dead end (German Edition) (pp.72-73). FISCHER E-Books. Kindle version.

> *"Unlike light microscopes, which use mirrors and lenses, particle accelerators use electric and magnetic fields to accelerate and focus beams of electrically charged particles. But by accelerating particles more and more to study an object, it*

becomes increasingly difficult to extract information from the measurement. This is because the particles that are supposed to examine the object begin to change it"(Translate: DeepL pro).

Kremer, C./ S. Aslanidou/ S. Bassauer/ M. Hilcker/ A. Krugmann/ P. von Neumann-Cosel/ T. Otsuka/ N. Pietralla/ V. Yu. Ponomarev/ N. Shimizu/ M. Singer/ G. Steinhilber/ T. Togashi/ Y. Tsunoda/ V. Werner/ M. Zweidinger. https://journals.aps.org/prl/abstract/10.1103/PhysRevLett.117.172503

Mayer-Kuckuk, Theo (1992): nuclear physics; ch. 6, nuclear models. B.G. Teubner Stuttgart 1992. (Teubner Studienbücher: Physik) 5th revised edition.
Otsuka T. /T. Abe / T. Yoshida/ Y. Tsunoda/ N. Shimizu/ N. Itagaki/ Y. Utsuno / J. Vary /, P. Maris / H. Ueno. (2022): α-Clustering in atomic nuclei from first principles with statistical learning and the Hoyle state character. Nat Commun 13, 2234 https://doi.org/10.1038/s41467-022-29582-0

"Radon." In: Wikipedia - The free encyclopedia. Edit date: March 17, 2023, 08:05 UTC. URL: https://de.wikipedia.org/w/index.php?title=Radon&oldid=231897223 (retrieved: 27 March 2023, 19:02 UTC).

Schmidt. A. et al; CLAS Collaboration (2020): Probing the core of the strong nuclear interaction. Article. Published: 2020/02/26. in: Nature, vol. 578, pages 540-544. https://www.nature.com/articles/s41586-020-2021-6

12.0 Abstract

The paper compares the scientific ideas of spherical and deformed atomic nuclei with the previously unknown theory of checkerboard and planar atomic structure. For this purpose, the interpretations of the nuclear structure of international researchers were compared with the author's interpretations according to the checkerboard and planar atomic structure.

This was done by presenting concrete investigation results of research teams and their interpretations and then commenting on them based on the opposite position. Several figures are provided for better understanding. It became clear that all scientific interpretations of the research reports are more or less based on the nuclear shell theory and the alpha particle cluster theory. Theories that do not provide convincing ideas about the core structure, even though they contain quite reasonable theoretical components.

However, it could be clarified that the counterposition of the checkerboard and planar atomic structure represents a consistent theory with which atomic nuclei can be represented and calculated. Here, the nuclear shapes arise from the nuclear structure by protons and neutrons and not vice versa. Atoms are built like planar gears because of the opposite rotation of protons and neutrons. In this respect, all atomic nuclei must be "flat".

Original basis of this theory are earlier research results, which prove the spin of proton and neutron and the same size and mass of the nucleons. This fact forms the basis for the nuclear structure of all atoms. Thus, nuclear shapes and nuclear structure can be explained conclusively and without contradiction.

Due to the extent of the topic some scientific theories and conceptions, in this comparison not or only partly could be considered.

13.0 Biography

Helmut Albert

Born 1960 in Speyer / Rhine

Lives and works in Freiburg / Germany

1974-1976 Apprenticeship as a locksmith / BASF Ludwigshafen.

1978-1981 Apprenticeship in the art trade

Evening Academy Mannheim: Nude drawing and art history.

1981-1987 Studies of sculpture at the Academy of Fine Arts, Nuremberg

(Works after the human figure. Material studies.

Works with steel, wood, plaster, rubber).

1987-2006 Concrete works based on material studies.

Since 2015 engagement with fundamental physics.

Awards:

2013 Artur Fischer Inventor Award of the State of Baden-Württemberg.

1995 One-year scholarship for fine arts, Offenburg.

Various art awards

2D ATOMS. Helmut M. Albert

FSC
www.fsc.org
MIX
Papier aus verantwortungsvollen Quellen
Paper from responsible sources
FSC® C105338